100 MIND-BLOWING ENGINEERING FEATS

100 Jaw-Dropping Innovations That Defy Limits

FELIX GRAYSON

MINDSPARK
PUBLISHING

CONTENTS

BEFORE WE DIVE IN...

Did you know that this is just **one** of many **mind-blowing** books waiting to be discovered?

What if I told you there's a **world of jaw-dropping, unbelievable, and downright bizarre facts** across **sports, science, history, mysteries, and more**—each one packed with stories that will **challenge what you thought you knew?**

EVER WONDERED WHAT IT'S LIKE TO...

- Witness **record-breaking Olympic moments** that defy human limits?

- Explore **real-life conspiracy theories** that sound too wild to be true?

- Discover **unsolved mysteries** that still leave experts baffled?

- Learn about **billionaires, stock market crashes, and money secrets?**

- Find out how **robots, AI, and space travel are shaping the future?**

- Experience the **most extreme sports, legend-ary battles, and shocking events?**

This is just the beginning. The **100 Mind-Blow-ing series** covers it **all.**

WANT TO SEE WHAT'S NEXT?

Go to **FelixGrayson.com** and explore the **grow-ing collection** of books and audiobooks that will **entertain, amaze, and keep you coming back for more.**

Curiosity doesn't stop here—this is just the beginning. What will blow your mind next?

INTRODUCTION

Welcome to *100 Mind-Blowing Engineering Feats*, a collection designed to make you say, "Wait… humans actually built that?" From gravity-defying skyscrapers to underwater tunnels and self-healing roads, this book is packed with jaw-dropping creations that push the limits of what we thought was possible.

Ever heard of a bridge that rolls itself into a ball? Or a power plant that doubles as a ski slope? How about a robot swarm that builds walls like a colony of ants—or a tower that pulls drinking water straight out of the sky? These are just a few of the mind-blowing feats waiting for you inside. Each one has been carefully chosen to inspire awe, spark curiosity, and maybe even make you look twice at the world around you.

Whether you're a curious reader, a future engineer, or just someone who loves a good "How did they do that?" moment, this book has something for you. Read it straight through, flip to a random page, or share your favorite facts with friends—there's no wrong way to enjoy the wild, wonderful world of engineering.

So grab a comfy seat, fire up your imagination, and get ready to explore 100 of the most incredible things humans have ever built. Let's dive in!

Mind-Blowing Engineering Feat #1

THE ISLAND THAT ENGINEERS BUILT

The entire country of the Netherlands is an engineering marvel—but perhaps nothing is more jaw-dropping than Flevoland: an entire province built *from the sea*.

In the mid-20th century, Dutch engineers undertook one of the boldest land reclamation projects in human history. Using massive dikes, pumps, and polders, they drained parts of the Zuiderzee—an inland sea—and transformed it into usable land. This was no small patch of dirt either: they created over **1,000 square kilometers** of entirely new land, big enough to fit a city the size of Los Angeles.

Today, Flevoland is home to over 400,000 people, complete with towns, roads, farms, and forests—all sitting below sea level. It's living proof that when nature says "no," engineers can still say, "watch this."

Mind-Blowing Engineering Feat #2

THE BRIDGE THAT BUILT ITSELF

The Millau Viaduct in southern France isn't just the tallest bridge in the world—it's also one of the most ingeniously constructed. Towering at **1,104 feet** at its highest point (even taller than the Eiffel Tower!), it spans the Tarn Valley with breathtaking elegance. But what makes this engineering feat truly mind-blowing is *how* they built it.

Instead of building the bridge from the ground up like most traditional bridges, engineers launched the roadway **horizontally** from each end, inching it out over the valley until the halves met in the middle. Think of it like sliding two decks of cards toward each other—except each "card" weighed thousands of tons!

This method minimized risk, reduced environmental impact, and cut down construction time. The result? A gravity-defying masterpiece that looks like it's floating in midair—and proves that sometimes the best way to build *up* is to build *out*.

Mind-Blowing Engineering Feat #3

THE TUNNEL BENEATH THE OCEAN

The Channel Tunnel—also known as the "Chunnel"—connects England and France with **a 31-mile railway tunnel under the English Channel**, making it one of the longest underwater tunnels ever built.

Constructed in the 1990s, the project required boring through layers of rock, chalk, and clay beneath the sea floor—all while avoiding flooding, shifting earth, and international tensions. At its deepest point, the tunnel runs **250 feet below the sea bed**. It took over **13,000 workers** and **six massive tunnel boring machines**—each the length of two football fields—to carve it out.

Today, high-speed trains whisk passengers between London and Paris in just over two hours, through a tunnel many once thought was impossible to build. What was once a treacherous sea crossing is now a smooth ride beneath the waves—thanks to some of the most daring engineering ever attempted.

Mind-Blowing Engineering Feat #4

THE DAM THAT REDIRECTED A RIVER

China's Three Gorges Dam isn't just the largest hydroelectric power station in the world—it literally reshaped the geography of a nation.

Spanning the Yangtze River, this colossal structure stretches 1.4 miles across and stands 607 feet tall. To build it, engineers had to do the unthinkable: **divert the third-longest river in the world.** That meant constructing massive cofferdams to hold back the Yangtze's flow during excavation and concrete pouring—an operation so massive it could be seen from space.

The dam contains over 28 million cubic meters of concrete, and its turbines generate enough electricity to power 80 million homes. It even slowed the Earth's rotation slightly due to the sheer volume of water it holds back.

Whether praised as a triumph or criticized for its environmental cost, the Three Gorges Dam remains one of the most staggering feats of modern engineering—proof that humans can move mountains, or in this case, rivers.

Mind-Blowing Engineering Feat #5

THE HOTEL BUILT IN 15 DAYS

In 2010, Chinese engineers shocked the world by constructing a **15-story hotel in just 6 days** — and then finishing the entire building in **15 days total**. The Ark Hotel, located in Changsha, was assembled using a revolutionary **modular construction method** that allowed workers to build entire floors from pre-fabricated units like giant LEGO pieces.

What's even more impressive? Despite the speed, the building met **strict earthquake safety standards** (designed to withstand a magnitude 9.0 quake), featured advanced insulation for energy efficiency, and produced **almost zero construction waste**.

While traditional skyscrapers often take years to complete, this project proved that with smart planning, innovation, and some jaw-dropping precision, a modern high-rise can quite literally go up overnight.

Mind-Blowing Engineering Feat #6

THE SHIP THAT SAILED ON LAND

When the **NASA Crawler-Transporter** was first built in the 1960s, it became the largest self-powered land vehicle on Earth—and it was designed for one incredible job: **carrying rockets**.

Weighing in at over **6 million pounds**, this behemoth was built to transport massive Saturn V rockets and later Space Shuttles from the Vehicle Assembly Building to the launch pad at Cape Canaveral. But here's the kicker: it only moves at **1 mile per hour** fully loaded, and yet it does so with **surgical precision**, maintaining balance and even leveling itself to keep the rocket perfectly upright over uneven terrain.

Its diesel engines generate over **2,750 horsepower**, and the crawler rides on **eight enormous tracks**, each made up of 57 shoes—each shoe weighing a ton. It's less a vehicle and more a mobile launch platform—an engineering feat that makes Earth itself feel like part of the launch process.

Mind-Blowing Engineering Feat #7

THE SKYSCRAPER THAT BREATHES

The **Burj Khalifa** in Dubai isn't just the tallest building in the world—it's a vertical city that *breathes*, *sways*, and *survives* in one of the harshest climates on Earth.

Standing at a staggering **2,717 feet**, the Burj Khalifa required revolutionary design solutions to handle fierce desert heat, brutal wind forces, and the challenges of building so high it nearly touches the clouds. Its sleek, spiraling design isn't just for looks—it's aerodynamic, breaking up strong winds that could otherwise destabilize the tower. And yes, it *moves*—up to **5 feet side-to-side** at the very top!

The tower also "breathes" through a condensation collection system that captures **15 million gallons** of water annually from the humid Gulf air—enough to irrigate its landscaping year-round.

From its soaring spire to its intelligent systems, the Burj Khalifa is more than just a building—it's a living, breathing feat of engineering mastery.

Mind-Blowing Engineering Feat #8

THE ROAD THAT MELTS ITSELF

In the snowy city of Reykjavik, Iceland, engineers came up with a genius way to deal with icy roads and sidewalks: **they built a heating system** *under* **them.**

Using the country's abundant geothermal energy, the streets and walkways in key areas are embedded with **a network of hot water pipes** that circulate natural heat from deep underground. When snow falls, the warm surfaces melt it almost instantly—no plows, no salt, no shovels required.

This isn't just convenient—it's incredibly efficient and eco-friendly. The system prevents accidents, reduces wear on infrastructure, and takes full advantage of Iceland's clean energy. In a country where the ground literally steams, engineers turned the heat beneath their feet into a self-melting street grid.

Mind-Blowing Engineering Feat #9

THE TOWER THAT CLEANS THE AIR

In the heart of smog-heavy Xi'an, China, engineers built a **giant air purifier disguised as a tower**—and it actually works.

Standing at **over 200 feet tall**, this experimental structure is designed to combat pollution on a city-wide scale. It sucks in dirty air at its base, filters it through several layers of advanced cleaning systems powered by solar energy, and releases purified air through the top. Early results showed a **reduction of up to 15% in airborne particulates** in the surrounding area—an impressive feat for a single structure.

What makes this even more mind-blowing? It runs *passively*, using the natural rise of heated air to keep it moving. No loud fans, no massive power requirements—just solar-heated glass and smart airflow. It's a skyscraper-sized air filter and a bold vision for the future of urban living.

Mind-Blowing Engineering Feat #10

THE ICE HOTEL THAT REBUILDS ITSELF

Every winter in the tiny village of Jukkasjärvi, Sweden, engineers and artists come together to create the **world's first hotel made entirely of ice and snow**—and then let it melt away in the spring.

Known as the **ICEHOTEL**, this architectural wonder is rebuilt **from scratch each year** using ice blocks harvested from the nearby Torne River. The structure includes suites, a bar, a chapel, and even ice sculptures you can sleep beside—all kept at a crisp **23°F (-5°C)**. No nails, no wood, no steel—just ice, snow, and engineering precision.

It takes about **5,000 tons of ice and snow** and a carefully timed construction process to ensure the structure stays stable and safe for guests. Then, when warmer temperatures return, the hotel melts back into the river it came from—leaving behind nothing but memories and blueprints for next year's frozen masterpiece.

Mind-Blowing Engineering Feat #11

THE WALL THAT HELD BACK A SEA

In Japan, the **Kamaishi Seawall** was once the *deepest breakwater ever built*, designed to shield coastal cities from devastating tsunamis.

Stretching **1.2 miles** into the Pacific and plunging to a depth of **207 feet**, this massive underwater wall was anchored to the seafloor with **concrete blocks weighing up to 8 tons each**. It took nearly **three decades** to complete and was meant to protect the industrial port of Kamaishi from nature's fury.

While it famously failed during the historic 2011 tsunami—overwhelmed by waves more powerful than anyone had imagined—it still slowed the incoming water just enough to **buy crucial seconds** that helped save lives. Engineers later studied its performance to improve future defenses, turning a partial failure into a foundation for innovation.

Sometimes, engineering feats don't just build— they *teach*, too.

Mind-Blowing Engineering Feat #12

THE FLOATING CITY THAT DOESN'T SINK

In the middle of the Pacific Ocean lies an engineering marvel that defies storms, waves, and even rising seas: the **U.S. Navy's Floating Instrument Platform**, or **FLIP**.

At first glance, FLIP looks like a weird, narrow ship. But when it reaches its destination, it does something no ship does—it **flips upright**, turning from a horizontal barge into a 355-foot vertical research tower floating in the ocean. Once in position, 300 feet of the platform extends below the surface, giving it **incredible stability**, even in rough seas.

Originally designed in the 1960s to study underwater acoustics, FLIP has helped scientists research everything from whale calls to wave patterns. It has **no engine, no propellers**, and needs to be towed into place—but once there, it becomes one of the most stable floating platforms ever engineered.

It's part submarine, part science lab, and 100% brilliant.

Mind-Blowing Engineering Feat #13

THE AIRPORT BUILT ON WATER

When space ran out in Osaka, Japan, engineers didn't expand inward—they built **an entire airport in the sea**.

Kansai International Airport sits on an *artificial island* in Osaka Bay, constructed from scratch on a foundation of **millions of tons of rock and sand**. The island is **2.5 miles long**, large enough for multiple runways, terminals, and even a train station. To protect it from typhoons and earthquakes, engineers designed it with deep-sea retaining walls and shock-absorbing foundations.

The biggest challenge? **It's slowly sinking.** Engineers anticipated this and designed the airport with adjustable supports and a flexible infrastructure that can shift as needed. Despite the ongoing battle against gravity and the sea, the airport has operated smoothly since opening in 1994—and stands as one of the most ambitious examples of reclaiming space where none existed.

Mind-Blowing Engineering Feat #14

THE POWER PLANT INSIDE A MOUNTAIN

Deep beneath a mountain in the Swiss Alps hides one of the most impressive—and in-visible—engineering feats in Europe: the **Nant de Drance pumped-storage power plant**.

Rather than build upward or outward, engi-neers dug **2,000 feet into the earth**, carving out enormous underground chambers to house **six powerful turbines**. The plant operates like a giant water battery: during low-demand hours, it pumps water uphill into a reservoir; during high-demand hours, it releases that water to generate electricity. It can deliver **900 megawatts** of power in min-utes—enough to stabilize an entire national grid.

Everything—from the generators to the tun-nels—is hidden within the mountain, preserv-ing the pristine Alpine landscape above. It's a mind-blowing blend of raw power and environ-mental stealth: an energy giant that you'd never know was there.

Mind-Blowing Engineering Feat #15

THE RAILWAY THAT CLIMBS A CLIFF

In the mountains of Peru, the **Central Railway of the Andes** pulls off one of the most daring climbs in railroad history—scaling the Andes to reach altitudes that leave passengers **gasping for air**.

Completed in the early 20th century, this railway ascends to over **15,800 feet above sea level**, making it one of the **highest railways in the world**. To conquer the steep terrain, engineers designed a labyrinth of **switchbacks, zigzags, and tunnels**—including the Galera Tunnel, which cuts through solid rock near the peak.

The construction faced brutal conditions: freezing temperatures, dizzying altitudes, and cliffs that drop off into nothingness. Workers had to blast through mountains and suspend track sections above ravines using sheer ingenuity and grit.

It's not just a railway—it's a gravity-defying ribbon of steel carved into the clouds.

Mind-Blowing Engineering Feat #16

THE ROBOT THAT BUILT A BRIDGE

In Amsterdam, engineers and designers teamed up to create something out of science fiction: a **3D-printed steel bridge**—built entirely by **robots**.

Using robotic arms equipped with welding tools, the project took place over several months, as the machines carefully laid down **molten metal layer by layer**, forming a fully functional pedestrian bridge. The final structure spans **40 feet**, weighs over **10 tons**, and features a futuristic, flowing design that would be nearly impossible to create with traditional methods.

What's more, the bridge is packed with **smart sensors** that track its usage and structural health in real time, feeding data back to researchers for future improvements.

It's not just a bridge—it's a blueprint for how robots might help shape the cities of tomorrow, one weld at a time.

Mind-Blowing Engineering Feat #17

THE PIPELINE THROUGH FROZEN EARTH

Running across **800 miles of Alaskan wilderness**, the **Trans-Alaska Pipeline System (TAPS)** is one of the most ambitious—and cold-defying—engineering feats in history.

Designed to transport oil from Prudhoe Bay in the Arctic to the southern port of Valdez, the pipeline had to overcome **permafrost**, earthquakes, wildlife, and extreme temperatures. Engineers elevated much of the pipe **above ground** on specially designed supports to prevent it from melting the frozen earth below—because if the permafrost thawed, the ground could collapse.

It includes **more than 78,000 vertical supports**, **hundreds of river crossings**, and even a **slide zone** where the pipe can flex and move if an earthquake hits. In one section, the pipe can shift more than **20 feet** without breaking.

It's a pipeline that doesn't just carry oil—it *dances* with nature to survive.

Mind-Blowing Engineering Feat #18

THE DESERT CITY THAT RECYCLES IT ALL

In the heart of the Arabian Desert, engineers are building a city so futuristic, it practically runs on **sustainability alone**. Welcome to **Masdar City**—an experimental urban hub in Abu Dhabi designed to produce **zero waste** and run entirely on **renewable energy**.

The city's layout was engineered to maximize shade and air circulation, dramatically reducing the need for air conditioning. Streets are narrow and walkable, and the entire city is powered by solar farms and clean tech. Even the **waste is recycled or reused**, from water to building materials.

One of the coolest features? An underground network of **autonomous electric pods** silently shuttles people around—no traditional cars allowed. Masdar is still growing, but even in its early stages, it's proving that cities of the future don't have to sacrifice comfort for sustainability.

It's not just a city—it's a working prototype of tomorrow.

Mind-Blowing Engineering Feat #19

THE BUILDING THAT ROTATES ITSELF

In Dubai, where bold architecture is the norm, one proposed skyscraper takes innovation to a whole new level—literally. The **Dynamic Tower** is designed to be the **world's first rotating skyscraper**, where each floor can spin **independently**.

Each of the tower's **80 floors** would be mounted on a central core, allowing residents to **control their apartment's rotation** with the push of a button. Want a sunrise view in the morning and a sunset panorama at night? Just spin your floor. The entire building would be powered by **wind turbines and solar panels**, making it not only mobile—but also sustainable.

Though not yet completed, the engineering plans are solid, and prototypes have proven the concept is feasible. It's a skyscraper that doesn't just scrape the sky—it *spins* through it.

Mind-Blowing Engineering Feat #20

THE SOLAR FARM THAT FLOATS

When land is scarce, engineers look to the water—and in China, they took that literally by building **one of the world's largest floating solar farms** atop a **flooded coal mine**.

Located in Anhui province, this renewable energy marvel features **over 160,000 solar panels** floating on the surface of a man-made lake. Not only does it generate clean power for **tens of thousands of homes**, but the water beneath helps **cool the panels**, boosting their efficiency.

Floating the farm on a disused mine site also avoids land competition with agriculture or cities—a smart reuse of otherwise unusable space. And because the panels shade the water, they even help **reduce evaporation** in hot conditions.

It's a perfect example of turning a legacy of pollution into a powerhouse of clean energy—one panel, and one ripple, at a time.

Mind-Blowing Engineering Feat #21

THE LIBRARY WITH NO LIBRARIANS

In Helsinki, Finland, engineers and designers teamed up to create **Oodi**, a futuristic library where **technology does the shelving, sorting, and scanning**—and humans just enjoy the experience.

Opened in 2018, Oodi is more than just a building full of books. It features **robotic book sorters**, **automated storage systems**, and self-service stations that allow visitors to check out materials without staff intervention. Behind the scenes, a system of **conveyor belts and robots** whisks returned books to their proper shelves—*faster than any librarian could*.

But Oodi isn't just about automation—it's about **redefining public space**. The building includes soundproof music studios, 3D printers, movie screening rooms, and even gaming zones, all powered by cutting-edge sustainable architecture.

It's not just a library—it's a tech-powered hub of creativity that quietly engineered the future of public learning.

Mind-Blowing Engineering Feat #22

THE TOWER THAT COOLS WITHOUT AC

In the sweltering heat of Harare, Zimbabwe, architects and engineers created a **28-story office building** that stays cool **without a single air conditioner**. Meet the **Eastgate Centre**, inspired by — of all things — **termite mounds**.

Termites build towering structures with natural ventilation systems that regulate temperature perfectly, even in extreme climates. Engineers mimicked this by designing Eastgate with **strategic vents, chimneys, and airflow channels** that pull in cool air at night and release warm air during the day. The result? A building that uses **90% less energy** for climate control than a conventional office tower.

It's a stunning example of **biomimicry** — using nature's blueprints to solve human problems. And in this case, it's proof that sometimes, the coolest solutions don't require a plug at all.

Mind-Blowing Engineering Feat #23

THE STADIUM THAT OPENS LIKE AN EYE

When engineers designed the **Al Janoub Stadium** in Al Wakrah, Qatar, they didn't just build a sports venue—they created a **mechanical marvel inspired by a sailboat**.

The stadium's most mind-blowing feature? Its **retractable roof**, which opens and closes **like a giant eyelid**. Made of ultra-lightweight materials stretched across a network of steel cables, the roof can transform the entire stadium in minutes— shielding fans from the desert heat or opening up to the night sky.

The design, inspired by traditional **dhow boats**, also includes a state-of-the-art **cooling system** that targets individual seating areas, reducing energy use in a massive open-air arena. With its futuristic curves and moving parts, Al Janoub isn't just built to host matches—it's engineered to perform.

Mind-Blowing Engineering Feat #24

THE LAB THAT LIVES AT THE SOUTH POLE

At the bottom of the world, where temperatures plunge below **-100°F** and winds howl across a frozen desert, stands one of the most extreme engineering feats on Earth: the **Amundsen–Scott South Pole Station**.

Perched atop **9,300 feet of ice**, this cutting-edge research facility had to be designed to survive brutal conditions *and* stay above the ever-accumulating snow. The solution? Engineers built it on **hydraulic stilts** that allow the entire station to be **jacked up and relocated**, preventing it from being buried.

The station houses scientists year-round, including during the six-month-long polar night. Inside, it's equipped with laboratories, living quarters, and observatories for studying everything from climate to cosmic rays—all while being **completely isolated from the rest of the planet**.

It's not just a building—it's a technological outpost on the edge of human endurance.

Mind-Blowing Engineering Feat #25

THE WALL THAT MOVES WITH THE EARTH

In earthquake-prone Los Angeles, engineers designed a high-rise so flexible, it's built to **move with the ground instead of resisting it**. The **Wilshire Grand Center**, standing **1,100 feet tall**, is not only the tallest building in California—it's also a masterclass in seismic design.

At its core is a **massive tuned mass damper**, a giant pendulum-like system that sways opposite the building's motion during an earthquake, reducing stress on the structure. The building's foundation includes a **record-breaking 21,200 cubic yards of concrete**, poured continuously over 18 hours to form a solid, flexible base.

The exterior is wrapped in a curtain wall system that can flex without shattering, and the internal steel frame is designed to bend—but not break—during seismic events.

It's a skyscraper that doesn't stand stiff against nature. It **dances with it**.

Mind-Blowing Engineering Feat #26

THE UNDERWATER ROUNDABOUT

In the Faroe Islands, a tiny North Atlantic archipelago, engineers built a road system so ambitious it dives **beneath the ocean floor**—complete with the world's **first underwater roundabout**.

Opened in 2020, the **Eysturoy Tunnel** stretches **over 7 miles**, connecting multiple islands via a deep-sea drive. But the real showstopper lies 600 feet below sea level: a **glowing, sculpted roundabout** carved into solid rock beneath the seabed. It directs traffic between three tunnel branches and features an illuminated art installation that turns navigation into a surreal experience.

Engineered to withstand immense pressure and built in one of Europe's harshest marine environments, the tunnel opened up faster travel, safer commutes, and even tourism for remote communities.

It's not just an intersection—it's a subterranean sculpture in motion.

Mind-Blowing Engineering Feat #27

THE FACTORY WITH NO WORKERS

In the city of Dongguan, China, engineers transformed an ordinary electronics factory into something extraordinary: a facility that runs almost entirely **without human workers**.

Dubbed the **"lights-out" factory**, this smart manufacturing plant uses **robotic arms, automated guided vehicles, and AI-powered systems** to build precision electronics around the clock—with **no need for lights, air conditioning, or breaks**. Since its automation overhaul, the factory cut labor by **90%**, increased output, and dramatically reduced defects.

Every part of the system—from inventory management to quality control—is handled by machines communicating in real time. Human technicians still monitor things remotely, but day-to-day operations happen without them on-site.

It's a glimpse into the future of manufacturing—where machines don't just assist, they *run the show*.

Mind-Blowing Engineering Feat #28

THE HIGHWAY THAT CHARGES CARS

On a test track in Sweden, engineers built a stretch of road that does something straight out of science fiction—it **charges electric vehicles as they drive**.

Known as **eRoadArlanda**, this pilot project features an embedded **electrified rail** that connects to a moving vehicle via a small arm beneath the car. As the car travels, it draws power directly from the road, recharging its battery on the go. Once the vehicle changes lanes or stops, the arm automatically retracts.

The road is smart, too—it tracks energy consumption per vehicle and bills accordingly. Designed to combat range anxiety and reduce the need for massive battery packs, this innovation could transform highways into **giant mobile charging stations**.

It's the electric road to the future—literally.

Mind-Blowing Engineering Feat #29

THE SKYSCRAPER THAT DRINKS THE FOG

In Lima, Peru—one of the driest cities on Earth—engineers and designers teamed up to create a **billboard that captures drinking water** straight from the air.

The structure, developed by engineers at UTEC (University of Engineering and Technology), looks like an ordinary roadside advertisement—but inside, it contains **dehumidifiers, filters, and storage tanks**. As coastal fog rolls in, the system pulls moisture from the air, purifies it, and stores it in a tap-accessible tank at the base.

In just three months, the billboard produced **over 2,500 gallons** of clean water—enough to supply local families who previously had little to no access.

It's not just a clever idea—it's a life-saving engineering solution disguised as an ad.

Mind-Blowing Engineering Feat #30

THE ROBOT THAT BUILDS ON THE MOON

Engineers at NASA and several international space agencies are developing robots that may soon **3D print buildings on the Moon**—using **lunar soil as the raw material**.

The concept relies on autonomous rovers equipped with **additive manufacturing tech**, designed to melt or bind **regolith** (the Moon's dusty surface) into solid structures. These machines could build landing pads, habitats, and storage units **before astronauts even arrive**—eliminating the need to haul tons of building materials from Earth.

What's more, these lunar structures would be **radiation-shielded, airtight, and thermally stable**, made with no human hands involved. The same technology could be adapted for Mars and other planets, laying the groundwork for off-world colonization.

It's not sci-fi anymore—engineers are literally preparing to **print civilization on the Moon**.

Mind-Blowing Engineering Feat #31

THE TRAIN THAT FLOATS ON AIR

In Japan, the **Maglev (magnetic levitation) train** doesn't ride on tracks—it **floats above them**, reaching speeds of over **370 mph** with zero physical contact.

The train uses **powerful superconducting magnets** to lift it several centimeters off the guideway, eliminating friction entirely. Propulsion comes from magnetic pulses that pull the train forward, allowing it to accelerate smoothly and silently through specialized tunnels and tracks. The result? A ride that's faster than most airplanes—and smoother than a luxury car.

In test runs, Japan's maglev has shattered speed records, and commercial lines are underway to link cities in record time. It's a technological leap that turns rail travel into something that feels more like flying.

It's not just a train—it's a **hovering bullet** slicing through the future.

Mind-Blowing Engineering Feat #32

THE DAM THAT STOPS SALTWATER

In the Netherlands, where land and sea are constantly at war, engineers built a **massive storm surge barrier** that can **seal off an entire estuary** — on command.

Known as the **Eastern Scheldt Storm Surge Barrier**, this titanic structure stretches over **5 miles** and features **gigantic sluice gates** that remain open most of the time, allowing tides to flow naturally and protect the ecosystem. But when a storm surge threatens, the gates can be **lowered in just 75 minutes**, instantly turning the open estuary into a protective wall against the sea.

Each gate weighs thousands of tons and is operated with extreme precision by a network of hydraulic systems and weather monitoring tech. Completed in 1986, the barrier is so complex it's often called **"the eighth wonder of the world."**

It's a dam that knows when to hold the line — and when to let nature breathe.

Mind-Blowing Engineering Feat #33

THE LAB THAT TRAVELS AT 17,500 MPH

Orbiting Earth every 90 minutes, the **International Space Station (ISS)** is the most advanced—and fastest—laboratory ever built by humans.

Launched and assembled piece by piece in orbit, the ISS is a **multi-national engineering marvel** made up of **over 100 modules**, stretching the length of a football field. It houses astronauts who live and work in **microgravity**, conducting experiments that are impossible to perform on Earth—like studying how fire behaves in space or how the human body changes without gravity.

To stay in orbit, the ISS travels at a blistering **17,500 miles per hour**, completing nearly **16 orbits a day**. It's constantly resupplied and adjusted by visiting spacecraft, and its systems are controlled by a mix of onboard crews and ground engineers around the world.

It's not just a space station—it's **a floating city of science**, racing above our heads.

Mind-Blowing Engineering Feat #34

THE CANAL THAT LIFTS SHIPS UP MOUNTAINS

In central China, engineers are constructing one of the most ambitious water transport systems in history: the **Three Gorges Ship Lift**, part of a massive project to move vessels **vertically** past the iconic dam.

Unlike traditional locks that slowly fill and drain, this **giant mechanical elevator** lifts ships weighing up to **3,000 tons** nearly **370 feet** in just **40 minutes**—a task that previously took hours using a series of cascading locks. The system uses a balanced counterweight mechanism, similar to an elevator, but on a scale large enough to carry a **multi-story building**.

It's the largest and highest ship lift in the world and allows cargo and passenger ships to bypass the dam quickly, efficiently, and with far less environmental strain.

It's not just moving water—it's moving **mountains** of steel and cargo, one boat at a time.

Mind-Blowing Engineering Feat #35

THE BRIDGE THAT FOLDS LIKE ORIGAMI

In London, there's a pedestrian bridge that doesn't just open—it **rolls itself into a ball**.

The **Rolling Bridge**, designed by Thomas Heatherwick, spans a small section of the Grand Union Canal. Instead of swinging or lifting like traditional drawbridges, this one is made of **eight triangular sections** connected by hydraulic pistons. When it's time to let boats pass, the bridge slowly **curls up** into an octagonal shape, folding in on itself with stunning grace.

Once the waterway is clear, it unfurls back into a walkway in just a few minutes. It's not just functional—it's *mesmerizing*, turning a simple crossing into a kinetic sculpture.

It's a bridge, it's a transformer, and it's proof that engineering can be both smart *and* beautiful.

Mind-Blowing Engineering Feat #36

THE ICEBREAKER THAT SAILS THROUGH SOLID ICE

In the frozen Arctic, where ordinary ships would be crushed like tin cans, a special class of vessels known as **nuclear-powered icebreakers** carve through **ten-foot-thick ice** like it's slush.

Russia's **Arktika-class** icebreakers are the most powerful ever built. Powered by **two nuclear reactors**, they generate over **80,000 horsepower** and can plow continuously through **solid sea ice**—opening up shipping routes in regions previously considered inaccessible. These ships aren't just boats—they're floating power plants with reinforced hulls designed to flex under pressure instead of crack.

They enable year-round access to the Northern Sea Route, drastically cutting travel time between Europe and Asia, and play a key role in Arctic research and exploration.

It's not just a ship—it's an **unstoppable force** designed to tame Earth's most hostile waters.

Mind-Blowing Engineering Feat #37

THE PARK BUILT ON AN OLD RAILROAD

In the heart of New York City, engineers and designers transformed a forgotten elevated railway into one of the most innovative urban green spaces in the world: the **High Line**.

Once an abandoned freight line over Manhattan's West Side, the structure was slated for demolition—until a team of visionaries turned it into a **1.5-mile-long park in the sky**. Engineers reinforced the aging steel infrastructure and integrated **native plants, walking paths, and public art**—all while preserving the original rails.

What makes the High Line so remarkable isn't just its beauty—it's how it reimagines **vertical space** in a dense city, proving that old infrastructure can be reborn into something vibrant, sustainable, and community-driven.

It's a garden, a gallery, and a sidewalk with a skyline view—all built on history.

Mind-Blowing Engineering Feat #38

THE MOUNTAIN CARVED BY LASER

In Norway, engineers used something far cooler than dynamite to carve through solid rock—they used **giant industrial lasers**.

During the construction of some of Norway's ultra-precise underground tunnels, traditional blasting wasn't an option due to proximity to delicate infrastructure. So engineers turned to **laser-guided rock cutting systems**, capable of vaporizing stone with **millimeter accuracy**.

These high-powered lasers, originally developed for mining and aerospace applications, create **smooth, stable tunnel walls** without vibrations or shockwaves—minimizing risk and reducing the need for additional support structures. The result is faster, safer tunneling in areas where every inch counts.

It's not the stuff of science fiction anymore—this is **rock surgery**, performed with beams of light.

Mind-Blowing Engineering Feat #39

THE CITY THAT MOVES ON TRACKS

In the vast deserts of Egypt, engineers pulled off something truly wild: they **moved an entire city's centerpiece**—*stone by stone*—to save it from being swallowed by water.

The ancient **Abu Simbel temples**, carved into rock over **3,000 years ago**, were directly in the path of rising waters from the newly constructed **Aswan High Dam**. Rather than let them drown, a global team of engineers and archaeologists **cut the temples into over 1,000 massive blocks**, some weighing **30 tons**, and **relocated them 200 feet higher** and **600 feet inland**.

The blocks were **precisely reassembled**, preserving the orientation so that sunlight still perfectly illuminates the inner sanctuary on the same two days each year as it did in ancient times.

It wasn't just preservation—it was an *engineering resurrection*.

Mind-Blowing Engineering Feat #40

THE ELEVATOR THAT GOES SIDEWAYS

In Germany, engineers revolutionized vertical transport with an elevator that doesn't just go **up and down**—it also goes **side to side**.

Called the **MULTI system**, this cable-free elevator uses **magnetic levitation**—the same tech behind maglev trains—to move **horizontally and vertically** through a building. Multiple elevator cabins can run in a continuous loop, reducing wait times and opening up entirely new architectural possibilities.

Tested in a 246-foot tall tower in Rottweil, this system eliminates the need for traditional elevator shafts, allowing buildings to be more space-efficient—and way more futuristic.

It's not just an elevator anymore—it's a **magnetic transport network**, wrapped in steel and ready to reshape the skyline.

Mind-Blowing Engineering Feat #41

THE SOLAR FURNACE THAT MELTS METAL

High in the French Pyrenees sits a structure that looks like something out of a sci-fi film—but it's very real, and it can reach temperatures over **5,400°F** using nothing but sunlight.

The **Odeillo Solar Furnace** is the **world's largest** solar-powered oven. It uses a massive array of **10,000 mirrors** to concentrate sunlight onto a single focal point, generating heat intense enough to **melt steel**, **vaporize materials**, and test spacecraft components.

No fuel. No flame. Just pure, focused solar energy harnessed with laser-like precision. Scientists use the furnace for research in materials science, thermodynamics, and even clean energy experiments.

It's the ultimate magnifying glass—an engineering marvel that turns **sunlight into firepower**.

Mind-Blowing Engineering Feat #42

THE TUNNEL THAT BREATHES FIRE

Beneath the Swiss Alps lies the **Gotthard Base Tunnel**—the **longest and deepest railway tunnel in the world**, stretching a staggering **35.5 miles** straight through solid rock.

But what truly sets it apart isn't just its length—it's the **fire-breathing safety system** engineered into every inch. In the event of a fire or emergency, the tunnel can **isolate heat and smoke**, switch airflow direction, and **blast air** through massive ventilation shafts to keep passengers safe. Specialized **fire-resistant escape routes** and **emergency trains** stand ready around the clock.

It took **over 2,400 workers** and nearly **two decades** to carve the tunnel through **mountains up to 7,500 feet high**, using a fleet of custom-built tunnel boring machines.

It's more than a passage through the Earth—it's a subterranean symphony of **precision, protection, and power**.

Mind-Blowing Engineering Feat #43

THE DOME THAT OPENS LIKE A CAMERA

In Atlanta, Georgia, engineers designed a stadium roof that doesn't just retract—it **unfolds like the iris of a camera lens**.

The **Mercedes-Benz Stadium** features a jaw-dropping **oculus-style roof** made of **eight massive triangular panels** that slide open and closed in a swirling, spiraling motion. The entire structure can fully open in about **10 minutes**, revealing the sky in dramatic fashion.

This isn't just a spectacle—it's a feat of synchronized engineering. Each panel weighs hundreds of tons and glides with such precision that the motion appears almost organic. The stadium is also LEED Platinum certified, packed with sustainable features and cutting-edge tech throughout.

It's not just a roof—it's **mechanical choreography** on a stadium-sized scale.

Mind-Blowing Engineering Feat #44

THE BUILDING THAT EATS SMOG

In Mexico City, where air pollution is a daily battle, engineers and architects created a building that doesn't just *resist* pollution—it **actively removes it from the air**.

The **Torre de Especialidades**, part of a local hospital, is clad in a futuristic skin of **"ProSolve370e" tiles**—a special type of **smog-eating facade**. These white, honeycomb-like panels are coated with **titanium dioxide**, which reacts with sunlight to **break down harmful air pollutants** into harmless substances.

One building, working passively, can neutralize the pollution from over **8,000 cars per day**—no moving parts, no power needed, just **sunlight and chemistry** doing the work.

It's more than a building—it's a **giant urban air purifier**, cleaning the sky one panel at a time.

Mind-Blowing Engineering Feat #45

THE ROAD THAT LIGHTS UP AT NIGHT

In a small town in the Netherlands, engineers created a bike path that **glows in the dark**, guiding riders with no need for streetlights.

Inspired by **Van Gogh's "Starry Night"**, the **Van Gogh-Roosegaarde Path** is embedded with **photoluminescent stones** that **absorb sunlight during the day** and **emit a soft glow at night**. The swirling patterns not only look magical but also enhance safety and visibility for nighttime cyclists.

No electricity, no wiring—just smart materials and artistic vision, fused into functional infrastructure. It's part of a larger push toward sustainable, interactive design that brings **engineering and creativity** together in everyday spaces.

It's not just a bike path—it's a **masterpiece you can ride on.**

Mind-Blowing Engineering Feat #46

THE POWER LINE THAT FLOATS ON ICE

In some of the coldest regions on Earth, engineers needed to run power lines across **frozen rivers and shifting glaciers**—so they built **floating transmission towers** on massive steel skids.

In remote parts of **Siberia and northern Canada**, these colossal towers don't have concrete foundations. Instead, they're mounted on **giant sled-like bases** that rest directly on the ice or permafrost. This design allows them to **move, flex, and be repositioned** as the terrain shifts seasonally—without collapsing or snapping the high-voltage cables they carry.

They're engineered to withstand **extreme winds, ice buildup, and temperatures below –60°F**, all while delivering electricity to some of the most inaccessible places on the planet.

They're not just towers—they're **frozen giants**, gliding across the tundra to keep the lights on.

Mind-Blowing Engineering Feat #47

THE STADIUM THAT DRINKS RAIN

In Singapore, where space is tight and sustainability is key, engineers built a stadium that **harvests rainwater** to help power itself.

The **Singapore National Stadium** features a massive domed roof—the largest of its kind—that not only shades spectators but also **funnels rainwater** into a collection system. This water is filtered and reused for **irrigation, cooling, and maintenance**, dramatically cutting the need for external water supplies.

Even the **bowl-shaped design** of the structure helps channel airflow, reducing the need for air conditioning despite the city's tropical heat. It's part of a greater effort to create buildings that don't just serve people—they work with nature.

It's a sports arena, a climate solution, and a **water-harvesting machine**, all rolled into one.

Mind-Blowing Engineering Feat #48

THE WALL THAT GROWS ITS OWN FOOD

In the middle of Milan, Italy, stands a skyscraper that doesn't just reach for the sky—it **feeds the people below**.

The **Bosco Verticale**, or "Vertical Forest," is a pair of residential towers covered in **over 20,000 trees, shrubs, and plants**—but it's more than a garden in the sky. Some of the vegetation includes **edible herbs and fruits**, allowing residents to **grow food on their balconies**, right outside their windows.

The greenery helps reduce noise, absorb CO_2, regulate temperature, and filter dust from the city air—all while creating a living ecosystem in the sky. Each tower is the equivalent of **20,000 square meters of forest** packed into a vertical footprint.

It's not just architecture—it's **urban farming,** wrapped around a high-rise.

Mind-Blowing Engineering Feat #49

THE CLOCK THAT WILL TICK FOR 10,000 YEARS

Deep inside a mountain in West Texas, engineers and inventors are building a timepiece unlike any other—the **10,000 Year Clock**.

Designed by the **Long Now Foundation** and funded in part by Amazon founder Jeff Bezos, this massive mechanical clock is being assembled **500 feet underground**. It's engineered to tick once a year, chime once a century, and strike a grand bell every **1,000 years**. No batteries. No external power. Just gears, gravity, and sunlight guiding its movements.

Every component is crafted from ultra-durable materials like **titanium and ceramic**, built to endure time, tectonics, and technology shifts. The idea? To inspire long-term thinking in a world obsessed with the short term.

It's not just a timekeeper—it's a **monument to deep time**, ticking toward the far future.

Mind-Blowing Engineering Feat #50

THE BRIDGE THAT GROWS ITSELF

In the dense jungles of Meghalaya, India, engineers didn't pour concrete or weld steel—they **trained living trees to build bridges**.

For centuries, the indigenous Khasi and Jaintia tribes have guided the roots of the **Ficus elastica** tree across rivers and ravines, weaving them around bamboo scaffolds until they take root on the other side. Over time, these **living root bridges** grow stronger, thicker, and more resilient—lasting **hundreds of years** and withstanding floods that would destroy man-made structures.

Some of these bridges stretch over **100 feet** and can support the weight of dozens of people at once. And the most astonishing part? They're **alive**—repairing themselves naturally and growing stronger with age.

It's engineering without machines—just **patience, nature, and generational wisdom**.

Mind-Blowing Engineering Feat #51

THE SKYSCRAPER THAT EATS EARTHQUAKES

In Taipei, Taiwan—an area prone to typhoons and earthquakes—engineers built **Taipei 101**, a supertall skyscraper with a secret weapon hiding inside: a **massive steel pendulum** that helps it stand tall.

Suspended between the 87th and 92nd floors, the **tuned mass damper** weighs **730 tons** and swings gently to counteract building sway during high winds or seismic activity. When the earth shakes or gusts slam into the tower, the damper moves in the opposite direction, keeping the structure stable.

This golden orb is so iconic it's visible to visitors—and even has its own mascot. But behind the charm is some of the most effective **passive seismic engineering** ever installed in a high-rise.

It's a skyscraper with a built-in **balancing act**, designed to dance with nature instead of fighting it.

Mind-Blowing Engineering Feat #52

THE UNDERSEA HOTEL SUITE

In the Maldives, luxury meets engineering in a way that's as bold as it is beautiful: **a hotel room built entirely beneath the ocean**.

Part of the **Conrad Maldives Rangali Island** resort, *The Muraka* is a two-level suite with a **fully submerged master bedroom**, enclosed in a 5-inch-thick acrylic dome. Guests sleep surrounded by coral reefs and tropical fish, with **180-degree panoramic views** of the Indian Ocean—**16 feet below the surface**.

Engineers had to design the structure to withstand intense underwater pressure, corrosion, and even tropical storms. The suite was prefabricated in Singapore, shipped in a custom dry dock, and gently submerged into place without disturbing the surrounding marine life.

It's not just a room—it's an underwater dream **built on precision and courage**.

Mind-Blowing Engineering Feat #53

THE MUSEUM THAT FLOATS ABOVE GROUND

In Mexico City, where the soil is soft and **prone to sinking**, engineers had a unique challenge: how do you build a massive museum on **unstable ground**? The answer: **don't let it touch the ground at all.**

The **Museo Soumaya**, with its iconic, shimmering, cloud-like shape, is supported by a **cantilevered steel frame** that distributes its weight to **just 28 contact points**—minimizing stress on the foundation. Over **16,000 hexagonal aluminum tiles** cover its curving façade, creating a structure that looks like it's floating in space.

Inside, there are **no columns** interrupting the open galleries—thanks to a complex spiral ramp and a **free-form design** that balances structural loads like a sculpture.

It's a museum, a landmark, and a masterclass in **elegant defiance of gravity**.

Mind-Blowing Engineering Feat #54

THE TOWER MADE OF WOOD

In an era dominated by steel and concrete, Norway took a bold leap backward—and forward—by building the **world's tallest wooden skyscraper**.

Standing **18 stories high**, the **Mjøstårnet** in Brumunddal is constructed almost entirely from **cross-laminated timber (CLT)**, a super-strong engineered wood that rivals steel in strength but is **lighter, renewable, and carbon-negative**. The structure includes apartments, office space, a hotel, and even a pool—proving that wood can do it all.

Engineers had to account for wind sway, fire safety, and long-term durability, using **advanced joinery and moisture control systems**. The result is a warm, modern tower that's also a **statement about sustainability and innovation**.

It's not a log cabin—it's a **timber titan** rewriting the rules of high-rise design.

Mind-Blowing Engineering Feat #55

THE WIND TURBINE TALLER THAN THE EIFFEL

O‌ff the coast of Scotland, engineers installed a wind turbine so massive it makes skyscrapers look small: the **Haliade-X**, one of the most powerful turbines ever built.

Standing at a jaw-dropping **853 feet tall**, with **blades longer than a football field**, each turbine can generate up to **14 megawatts** of electricity—enough to power **over 16,000 homes per unit**. Its massive rotor sweeps an area larger than **seven football fields**, capturing more wind and producing more energy than any turbine before it.

What's more, the Haliade-X is designed for offshore wind farms, where stronger and more consistent winds can deliver **clean energy at scale** without occupying valuable land.

It's not just a turbine—it's a **giant in the wind**, spinning us toward a greener future.

Mind-Blowing Engineering Feat #56

THE AQUARIUM WITH A 4-STORY GLASS WALL

In Osaka, Japan, the **Kaiyukan Aquarium** features a jaw-dropping centerpiece: a **9-meter-high acrylic viewing window** holding back **5,400 tons of water**—and housing **whale sharks**, the largest fish on Earth.

This four-story glass wall isn't made of regular glass at all—it's composed of **ultra-thick, curved acrylic panels** seamlessly bonded together to withstand **immense pressure** without distorting the view. The engineering challenge? Make it **transparent, strong, and flawless**, so visitors feel like they're standing inside the ocean.

To support the structure, engineers used precise pressure modeling and a custom steel frame that flexes ever so slightly, allowing the panel to absorb vibrations and slight shifts without cracking.

It's more than an exhibit—it's an **underwater cathedral** built on clarity, courage, and craftsmanship.

Mind-Blowing Engineering Feat #57

THE BUILDING THAT MOVES WITH THE SUN

In Hyderabad, India, engineers created an office building that **rotates with the sun**, reducing energy use while dazzling the skyline.

Called the **Cybertecture Egg**, this futuristic structure is mounted on a **central core** and rotates slowly throughout the day to **maximize natural light** and **minimize heat gain**. Its curved, egg-like shape also reduces wind resistance and increases structural efficiency—using **less material** than a conventional building of the same size.

The façade is covered with **smart glass** that adjusts tint based on sunlight, and the building uses **greywater recycling,** passive cooling systems, and rooftop solar panels to stay as green as it is futuristic.

It's an office that **turns with the Earth**, engineered for harmony with the elements.

Mind-Blowing Engineering Feat #58

THE HIGHWAY THAT RIDES THE WAVES

In Japan, where land is scarce and earthquakes are frequent, engineers built a **floating expressway** over Tokyo Bay—designed to **withstand tsunamis, storms, and seismic shocks**.

The **Tokyo Bay Aqua-Line** is a hybrid mega-project: part bridge, part **underwater tunnel**, and part **floating roadway**. The floating sections are supported by **hollow concrete pontoons** anchored to the seabed, allowing them to **move flexibly with tides and waves** without cracking or collapsing.

The system also includes high-tech sensors to monitor structural health, and emergency shelters built into rest areas above the water. Even the transition from floating road to tunnel was engineered with **shock-absorbing joints** that can flex during earthquakes.

It's not just a highway—it's a **storm-surfing superstructure**, gliding above and below the sea.

Mind-Blowing Engineering Feat #59

THE DATA CENTER COOLED BY THE SEA

To cut down on energy use and rethink digital infrastructure, Microsoft sunk a **fully operational data center** to the bottom of the **North Sea**.

Called **Project Natick**, this underwater data center housed **864 servers** and operated for **two years** beneath the waves off Scotland's coast. The cold ocean water acted as a **natural coolant**, drastically reducing the energy needed to keep the servers from overheating—one of the biggest challenges in traditional data centers.

Engineers also discovered the sealed, oxygen-free environment led to **far fewer hardware failures** compared to land-based setups. Once retrieved, the system was fully intact and provided valuable insights into sustainable computing.

It's not just cloud storage—it's **deep-sea data**, engineered to ride the currents of the digital age.

Mind-Blowing Engineering Feat #60

THE FACTORY THAT FITS IN A BOX

In remote or disaster-stricken areas, building medical supplies can be nearly impossible—so engineers developed a **portable factory** that fits inside a **shipping container**.

Created by researchers at MIT, these compact, mobile units are capable of **producing thousands of vaccine doses per day**, complete with sterilization systems, quality control tech, and remote monitoring tools. The entire setup can be **shipped anywhere in the world**, deployed quickly, and powered using local infrastructure or generators.

Each container is a **fully functional pharmaceutical lab**, designed to respond rapidly to outbreaks or support regions without access to large-scale manufacturing facilities.

It's not just logistics—it's **life-saving science**, packed into a box and ready to roll.

Mind-Blowing Engineering Feat #61

THE WALL THAT CHANGES WITH THE WIND

In Sydney, Australia, engineers gave an ordinary building an extraordinary upgrade—a **living, moving facade** that **reacts to the wind**.

The **One Central Park** complex features a vertical garden covering its surface, but the real marvel is its **heliostat system**: a series of **motorized mirrors** mounted on a **cantilevered panel** above the tower. These mirrors rotate throughout the day to **bounce sunlight** deep into shaded areas of the building and surrounding park.

To complement this, parts of the building's facade are equipped with **kinetic panels** that flutter and shift with the breeze, reducing heat and glare while creating a shimmering, dynamic texture across the structure.

It's not just architecture—it's a **building that breathes, bends, and beams** with the environment.

Mind-Blowing Engineering Feat #62

THE ISLAND THAT MAKES ITS OWN LAND

In Dubai, engineers didn't just build structures—they **built an island chain** from scratch, reshaping the Persian Gulf with **palm trees and precision.**

The **Palm Jumeirah** is a sprawling, palm-shaped artificial island made of **over 100 million cubic meters of sand and rock**, dredged from the sea and placed with satellite-guided accuracy. No concrete foundation—just **geo-engineered seabed** reinforced with **geotextiles** and layers of protective rock to withstand waves and erosion.

Constructing the palm's intricate fronds and crescent breakwater required **custom dredging ships**, advanced hydraulic modeling, and nonstop environmental monitoring to minimize impact and maintain stability.

It's not just a luxury resort—it's a **geoengineering spectacle**, built where no land existed before.

Mind-Blowing Engineering Feat #63

THE CHURCH BUILT INTO A VOLCANO

In Lalibela, Ethiopia, engineers in the 12th century carved an entire complex of churches **downward into volcanic rock**—not built from stone, but **hewn directly out of it**.

The most iconic of these, the **Church of Saint George**, was carved from a single solid block of red basalt, forming a perfectly cross-shaped structure that descends **40 feet below ground level**. No scaffolding, no cranes—just **chisels, hammers, and centuries-old ingenuity**.

The churches are connected by a maze of tunnels, trenches, and passageways, forming a sunken sanctuary that has stood for over 800 years, resistant to weather, war, and time.

It's not just a place of worship—it's a **monolithic masterpiece**, engineered straight from the Earth itself.

Mind-Blowing Engineering Feat #64

THE BRIDGE THAT GROWS CORAL

Off the coast of Bali, engineers and marine biologists joined forces to build a **bridge that repairs coral reefs**—using **3D-printed, reef-friendly concrete**.

The **Living Seawall Project** combines infrastructure and ecology by embedding specially shaped concrete panels into bridge foundations. These panels mimic the **texture and complexity of natural coral**, providing a perfect home for marine life to attach, grow, and thrive.

As tides flow through the structure, nutrients and larvae settle in the crevices, turning a static bridge support into a **living reef system**. Over time, the structure strengthens both **biodiversity and coastal resilience**, helping reduce erosion and restore damaged habitats.

It's a bridge that **doesn't just connect land—it heals the ocean**.

Mind-Blowing Engineering Feat #65

THE TOWER THAT HARVESTS LIGHTNING

In a bold blend of architecture and atmospheric science, engineers in São Paulo, Brazil, designed a skyscraper that doesn't just withstand lightning—it **collects data from it**.

The **Millennium Tower** is outfitted with a special **lightning observation system** built directly into its lightning rod. When a strike occurs, sensors capture **high-speed electrical and magnetic field data**, helping researchers understand the behavior of lightning in real time.

This data is then used to improve lightning protection systems, aircraft safety, and even early warning systems for extreme weather events. The tower has become one of the **most studied lightning targets in the world**, struck dozens of times per year.

It's not just resisting the storm—it's **learning from every bolt that hits**.

Mind-Blowing Engineering Feat #66

THE GREENHOUSE THAT WORKS IN THE ARCTIC

Near the Arctic Circle, where sunlight is scarce and farming seems impossible, engineers built a **self-sustaining greenhouse** that grows fresh food year-round—with no soil and almost no sunlight.

Located in Nunavut, Canada, the **Growing North Project** uses a **geodesic dome greenhouse** equipped with **hydroponic systems, LED grow lights, and solar panels**. The dome's shape maximizes thermal efficiency, retaining heat even in **–40°F temperatures**, while the controlled interior grows crops like lettuce, herbs, and tomatoes—right in the tundra.

The entire system is optimized for **off-grid use**, helping remote Indigenous communities gain food security, lower grocery costs, and reduce carbon footprints from imported goods.

It's not just farming—it's **engineering photosynthesis in the polar night.**

Mind-Blowing Engineering Feat #67

THE CRANE THAT CLIMBS ITSELF

When constructing supertall skyscrapers, one question always looms: **How do you lift a crane as the building rises?** Engineers answered with a jaw-dropping solution—a **crane that climbs itself**.

Used on towers like the **Burj Khalifa**, these **self-climbing tower cranes** are mounted inside the building's elevator shaft or structural core. As new floors are added, the crane uses **hydraulic jacks** to lift itself up to the next level—like a giant mechanical inchworm.

This allows construction to continue vertically **without disassembling and reassembling** the crane every few floors, saving time, space, and logistics complexity. It's one of the few machines on Earth that **builds the thing it stands on—while standing on it**.

It's not just a lifting device—it's a **sky-climbing marvel of mechanical choreography**.

Mind-Blowing Engineering Feat #68

THE TUNNEL THAT SPINS ITS WAY DOWN

In Norway's rugged mountains, engineers faced a wild problem: how do you get cars safely down a steep fjord cliff without destroying the landscape? The answer? **A spiral tunnel.**

The **Lærdal Tunnel System** includes the **Spiral Tunnel of Åkrafjord**, a mind-blowing design where the road **loops in a corkscrew** *inside the mountain itself*, gradually lowering drivers through **tight, circular paths** carved directly from solid rock.

These loops reduce the need for switchbacks or massive surface cuts and allow vehicles to descend steep terrain **safely and smoothly**—while completely hidden from view. The tunnel system also includes **mood lighting and rest zones** to combat driver fatigue during the descent.

It's not just a tunnel—it's a **drivable helix**, spiraling through stone.

Mind-Blowing Engineering Feat #69

THE DRONE THAT BUILDS LIKE A BEE

Inspired by nature, engineers have developed flying robots that don't just deliver packages — they **build structures in mid-air**, like swarms of futuristic bees.

Known as **"BuilDrones,"** these autonomous aerial robots were tested by researchers at Imperial College London and ETH Zurich. Working together, they can **3D print structures on-site**, layer by layer, using lightweight materials like foam and special cement — no scaffolding or human hands required.

Each drone is equipped with **vision systems and real-time coordination software**, allowing them to **adapt to wind, errors, or design changes** on the fly. They can reach hard-to-access areas or even repair infrastructure in dangerous environments.

It's not just construction — it's **skyborne teamwork**, buzzing with precision and possibility.

Mind-Blowing Engineering Feat #70

THE BRIDGE THAT ROLLS OUT LIKE A CARPET

In emergencies or military operations, engineers often need a bridge *right now*. Enter the **Mobile Folding Bridge**, a brilliant invention that **unrolls itself like a giant metal carpet**.

Developed by military engineers, these bridges are mounted on specialized vehicles and can be **deployed in under 2 minutes**. Once in place, they create an instant crossing over rivers, trenches, or broken roads—supporting the weight of tanks, trucks, or emergency responders.

The bridge folds and unfolds via **hydraulic arms and telescoping sections**, requiring no cranes, no assembly crews, and no foundation work. After use, it's simply reeled back in, like a colossal roll of steel ribbon.

It's not just portable infrastructure—it's **rapid response engineering on wheels**.

Mind-Blowing Engineering Feat #71

THE WINDOW THAT GENERATES POWER

What if every window in a skyscraper could act like a solar panel? Engineers are making it happen with **transparent solar glass**—a game-changing material that turns windows into **clean energy sources**.

Developed at universities and clean tech firms around the world, this special glass uses **organic photovoltaic cells** or **quantum dots** to capture **invisible infrared and ultraviolet light**, while still allowing visible light to pass through. That means buildings can **harvest sunlight** without sacrificing natural lighting or aesthetics.

Installed in facades, bus stops, greenhouses, and even phone screens, these solar windows generate electricity **without anyone noticing**—and without taking up extra space.

It's not just a view—it's a **power plant you can see through.**

Mind-Blowing Engineering Feat #72

THE AIRPORT BUILT ON A FLOATING RUNWAY

In Japan's Osaka Bay, where land is scarce and earthquakes are frequent, engineers pulled off a jaw-dropping feat: they built **an entire airport on an artificial island**—and gave it a runway that **floats**.

Kobe Airport sits on a manmade island supported by a unique system of **seismic dampers and floating slabs**, designed to absorb the shock of earthquakes and typhoons. Its runway and terminal rest on a bed of engineered fill and flexible supports that **rise and fall** with ground movement and sea conditions.

The entire airport was constructed offshore to minimize noise over the city and allow for **24-hour operations**—something impossible on the mainland. Engineers even accounted for **ongoing land settlement**, designing adjustable foundations that can be recalibrated over time.

It's not just an airport—it's a **floating fortress of flight**, engineered to defy both nature and gravity.

Mind-Blowing Engineering Feat #73

THE HOSPITAL THAT DRIVES ITSELF

In remote parts of China and Africa, engineers have deployed a groundbreaking solution to bring healthcare where roads barely exist: a **self-driving mobile hospital**.

These autonomous medical units are built into **AI-powered vehicles** equipped with **robotic arms, diagnostic scanners, solar panels**, and **telemedicine systems** that connect patients with doctors in real time. Some even include **onboard labs** and **vaccine storage**, making them fully functional mini-clinics on wheels.

They navigate tough terrain using **LiDAR, GPS, and obstacle-avoidance systems**, and can operate with **no driver and no local infrastructure**, bringing care to the most underserved corners of the planet.

It's not just mobility—it's **medicine in motion**, engineered to save lives anywhere on Earth.

Mind-Blowing Engineering Feat #74

THE SKYSCRAPER COOLED BY A LAKE

In Toronto, engineers found a genius way to cool dozens of high-rise buildings—**with water from the bottom of a lake.**

The **Deep Lake Water Cooling System** uses pipes that reach **83 meters (272 feet)** into **Lake Ontario**, where the water stays a chilly **39°F (4°C)** year-round. This naturally cold water is pumped into a heat exchange system that cools entire office towers, hospitals, and data centers—**without traditional air conditioning.**

The result? A system that uses **up to 90% less electricity**, dramatically reducing greenhouse gas emissions and energy costs. It's one of the largest and most efficient of its kind in the world.

It's not just cooling—it's **liquid engineering,** straight from nature's own icebox.

Mind-Blowing Engineering Feat #75

THE ROBOT THAT REPAIRS UNDERWATER PIPES

Beneath the ocean's surface, where visibility is low and pressure is high, engineers developed a robotic marvel that can **inspect and fix underwater pipelines**—*without surfacing once.*

Known as **"autonomous underwater vehicles" (AUVs)**, these torpedo-shaped robots are equipped with **sonar, cameras, manipulators, and AI**, allowing them to **scan, detect leaks, and perform maintenance** on vital oil and gas infrastructure thousands of feet below the surface.

Some models, like the **Saab Sabertooth**, can **dock at underwater charging stations** and stay submerged for months, returning only when needed. They reduce the need for costly human dive teams or large maintenance ships—and can work in **complete darkness and crushing pressure**.

It's not just a drone—it's a **subsea engineer**, built to fix the unseen.

Mind-Blowing Engineering Feat #76

THE POWER PLANT FUELED BY TRASH

In Copenhagen, engineers turned garbage into green gold by building **Amager Bakke**—a power plant that **burns waste to generate electricity and heat** for over 150,000 homes.

But this isn't just any waste-to-energy facility. Amager Bakke is also a **ski slope**. That's right—the roof of the plant is covered in artificial turf and equipped with **ski lifts**, hiking trails, and even a climbing wall, making it one of the **most fun-filled power plants on Earth**.

The plant uses cutting-edge filters and scrubbers to ensure **ultra-low emissions**, turning trash into energy with **minimal environmental impact**. It's part of Denmark's ambitious goal to become **carbon-neutral** while rethinking urban design.

It's not just a power plant—it's **trash-powered recreation**, engineered with flair.

Mind-Blowing Engineering Feat #77

THE SHIP THAT STANDS ITSELF UP

Off the coast of California, engineers designed a ship that **doesn't sail—it sinks and stands upright** in the water to become a floating science lab.

Meet the **FLIP (Floating Instrument Platform)**—a 355-foot-long vessel that transitions from horizontal to vertical by **flooding ballast tanks** in its stern. Once upright, **300 feet of the structure submerges**, making it incredibly stable in rough seas—**perfect for oceanographic research**.

FLIP has no engine and must be towed into place. But once deployed, it becomes a nearly motionless platform used to study **sound waves, currents, marine life, and more**—even during hurricanes.

It's not just a ship—it's **a standing sentinel of the sea**, engineered to flip the script on ocean science.

Mind-Blowing Engineering Feat #78

THE HOUSE THAT BUILDS ITSELF IN HOURS

In disaster zones or remote regions, engineers have created a housing solution that arrives flat-packed and **assembles itself in just a few hours**—with no tools required.

Called the **Ten Fold House**, this ingenious structure is a **mechanically unfolding building** that expands from a compact box into a fully functional living or office space using **built-in counterweights and hinges**. Just press a button, and the entire house unfolds—**walls, floors, roof, and all**—ready to use.

No cranes, no concrete, no construction crews. The modular design can be relocated, reused, and customized, offering rapid shelter in emergencies or flexible spaces in remote areas.

It's not just a house—it's **pop-up architecture**, engineered for speed, simplicity, and survival.

Mind-Blowing Engineering Feat #79

THE TUNNEL DUG BY A GIGANTIC WORM

When engineers in Seattle needed to dig a massive highway tunnel beneath the city, they unleashed **Bertha**—a tunnel boring machine so huge it looked like **a mechanical earthworm the size of a building**.

Stretching **57 feet in diameter** and weighing **7,000 tons**, Bertha was one of the **largest tunnel boring machines ever built**. It carved a 2-mile-long path under downtown Seattle, digging, reinforcing, and sealing the tunnel **as it moved forward**, inch by inch.

Bertha faced intense challenges—like overheating, breakdowns, and unexpected obstacles—but eventually completed her mission, creating a new underground highway while the city carried on above.

It's not just tunneling—it's **mega-digging on a scale that could eat cities**.

Mind-Blowing Engineering Feat #80

THE TRAIN THAT CLIMBS A WATERFALL

In Switzerland, where mountains dominate the landscape, engineers built the **Gelmerbahn**—a train so steep it feels more like a vertical ride than a railway.

Originally constructed to transport materials for a hydroelectric dam, the Gelmerbahn is now a **passenger funicular railway** that climbs a **106% gradient—the steepest open-air funicular in Europe**. The track ascends alongside cliffs and waterfalls, with no seatbelts, doors, or cabins—just **an open-air platform on rails**.

It relies on **a counterweight system** powered by gravity, where one car goes up while the other comes down, perfectly balanced and astonishingly smooth.

It's not just a train—it's **a mountain-climbing thrill ride**, engineered for the edge.

Mind-Blowing Engineering Feat #81

THE STADIUM THAT FLOATS ON WATER

In Singapore, where land is precious, engineers created a solution that redefined what a stadium could be: they **built it on the water**.

The **Float @ Marina Bay** is the **world's largest floating stage**, made of **steel pontoons** anchored in the bay and capable of supporting **more than 1,000 tons**—enough for full concerts, soccer matches, and military parades. It's paired with a **30,000-seat grandstand** on the shore, offering sweeping skyline views.

The modular platform can be **reconfigured or relocated**, and its buoyancy system is designed to withstand strong currents and monsoon rains. During national holidays, it doubles as a launch pad for **massive fireworks shows** over the bay.

It's not just a stadium—it's **a stage that sails**, engineered to perform on water.

Mind-Blowing Engineering Feat #82

THE BUILDING WRAPPED IN ALGAE

In Hamburg, Germany, engineers unveiled a world-first: a building powered and insulated by **living algae**.

Called the **BIQ House**, this experimental apartment complex features **bioreactor panels** filled with microalgae mounted on its exterior. These glass panels **absorb sunlight**, enabling the algae to grow while **generating biomass and heat**, which is then harvested to power the building's systems.

The algae also act as **dynamic shading devices**, adjusting naturally with sunlight to reduce interior heat gain. As the algae blooms and thickens, it filters light and enhances energy production—**a living, breathing facade** that adapts to the environment.

It's not just green design—it's **architecture that photosynthesizes.**

Mind-Blowing Engineering Feat #83

THE TOWER THAT DRINKS THE SKY

In the dry hills of Morocco, engineers built a tower that **pulls water out of thin air**—no pipes, pumps, or power needed.

Called the **Warka Tower**, this lightweight, 30-foot-tall structure uses **mesh netting and bamboo framing** to harvest moisture from the air through **condensation**. As humid air passes through the mesh, droplets form and trickle down into a **collection basin**, producing up to **26 gallons of clean water per day**.

Inspired by the **warka tree**, a symbol of shelter in Ethiopia, the tower is designed to be **easily assembled by local communities**, using affordable and sustainable materials.

It's not just a water collector—it's a **sky-harvesting beacon of hope**, engineered for survival in the harshest climates.

Mind-Blowing Engineering Feat #84

THE DOME THAT INFLATES IN MINUTES

In emergencies, sports, or even space missions, speed and portability matter—so engineers created **inflatable domes** that can rise from a backpack to a building in minutes.

Known as **air-supported structures**, these domes use **constant internal air pressure** to maintain their shape—**no beams, no columns, just fabric and fans**. Some, like the **Theia Dome**, can cover a full soccer field, while others are used as **temporary hospitals, disaster shelters, or event venues**.

The materials are tough—often **PVC-coated polyester**—and can withstand wind, rain, and even snow. Best of all, they're **ultralight, reusable, and deployable anywhere**, with minimal setup.

It's not just a tent—it's **architecture on demand**, engineered to rise with the air itself.

Mind-Blowing Engineering Feat #85

THE ICE ROAD THAT MELTS EVERY YEAR

In Canada's remote Northwest Territories, engineers created a highway that exists for only part of the year—because it's made entirely of **frozen water**.

Known as the **Tibbitt to Contwoyto Winter Road**, this seasonal route spans over **370 miles**, with **85% of it crossing frozen lakes**. Built each winter to supply diamond mines deep in the Arctic, the road is engineered by **clearing snow to thicken the ice** and **using radar and drills** to measure ice integrity daily.

Trucks weighing up to **40 tons** cross this icy highway—but only during a short window when temperatures are cold enough to keep the surface stable. When spring comes, the entire road **disappears**, leaving no trace behind.

It's not just a road—it's a **temporary lifeline**, engineered from ice and timing.

Mind-Blowing Engineering Feat #86

THE SOLAR PLANT THAT FOLLOWS THE SUN

In the Nevada desert, engineers built a power plant that doesn't just use sunlight—it **chases it across the sky**.

The **Crescent Dunes Solar Energy Project** uses **10,347 mirrors**, each the size of a garage door, that automatically rotate to track the sun throughout the day. These **heliostats** reflect sunlight onto a central **molten salt tower**, heating it to over **1,000°F** to generate steam and produce electricity—even **after sunset**.

The molten salt retains heat for hours, allowing the plant to deliver **round-the-clock renewable energy**, a huge step toward replacing fossil fuels with reliable solar power.

It's not just solar—it's **sun-tracking precision**, engineered to capture every ray.

Mind-Blowing Engineering Feat #87

THE ELEVATOR THAT CLIMBS INTO SPACE

It may sound like science fiction, but engineers around the world are actively developing a concept so bold it could change space travel forever: a **space elevator**.

The idea? Anchor a **tether to Earth** and stretch it all the way to **geostationary orbit**, using **carbon nanotubes or graphene**, materials **stronger than steel but lighter than plastic**. A motorized "climber" would travel up the tether, carrying cargo and potentially humans—**without rockets**.

Japan's Obayashi Corporation has outlined plans for a working model by 2050, and prototypes are already being tested in labs and in microgravity. The challenges are enormous, but the promise? **Low-cost, reusable access to space**—with less fuel and less risk.

It's not just a dream—it's **an elevator to the stars**, engineered to launch humanity skyward.

Mind-Blowing Engineering Feat #88

THE SKYSCRAPER THAT CATCHES RAIN

In drought-prone Nairobi, Kenya, engineers designed a high-rise that doesn't just scrape the sky—it **harvests it**.

The **Britam Tower**, one of Africa's tallest buildings, features an angular, sloped crown that funnels rainwater into a **massive underground reservoir**. This water is then **filtered, stored, and reused** throughout the building for plumbing, cooling, and irrigation—**reducing reliance on municipal supply**.

The design also integrates **natural ventilation**, **sunlight-maximizing angles**, and **high-efficiency glass**, cutting energy use dramatically in a hot, urban environment.

It's not just a tower—it's a **vertical oasis**, engineered to live off the clouds.

Mind-Blowing Engineering Feat #89

THE DAM THAT TURNS ITSELF TRANSPARENT

In China's Yunnan province, engineers constructed a dam with a high-tech twist—it can **vanish from view** during the day.

The **Baima Snow Mountain Dam** features a facade made from **mirrored stainless steel**, designed to reflect its surroundings so perfectly that it **blends into the landscape**. Against the snowy peaks and blue sky, the dam almost disappears, reducing its visual impact on the **UNESCO-protected nature reserve** nearby.

Behind the beauty lies function: it's a **hydropower facility** generating clean electricity, while its reflective surface also helps regulate heat and resist corrosion—preserving both aesthetics and performance.

It's not just infrastructure—it's **camouflaged engineering**, built to vanish into nature.

Mind-Blowing Engineering Feat #90

THE PRINTER THAT BUILDS HOUSES

In several parts of the world, engineers are now building homes not with hammers and nails — but with **giant 3D printers**.

Using a robotic arm and a nozzle that extrudes **special cement-like material**, these printers can construct the **walls of an entire house in under 24 hours**. The layers are laid down **precisely and rapidly**, reducing waste, labor costs, and build time — while allowing for **curved walls, insulation channels, and unique architectural designs** that traditional methods struggle to achieve.

Projects in Mexico, the U.S., and the Netherlands have already produced **entire communities** using this technology — some even off-grid and disaster-resistant.

It's not just a tool — it's **a blueprint for the future**, where homes are *printed, not poured.*

Mind-Blowing Engineering Feat #91

THE WALL THAT STOPS A GLACIER

In Peru, engineers are battling climate change with something extraordinary: a **man-made wall built to hold back a melting glacier.**

The **Lake Palcacocha** glacier above Huaraz is rapidly melting due to rising temperatures, threatening to unleash a deadly flood on the city below. To prevent catastrophe, engineers constructed a **reinforced dam and drainage system**—complete with **automated siphons and emergency spillways**—to **control the lake's rising waters**.

They also installed **early warning systems** and satellite monitoring to track changes in glacier volume in real time, giving residents critical minutes to evacuate if needed.

It's not just disaster prevention—it's **climate defense engineering**, holding back a wall of water from the sky.

Mind-Blowing Engineering Feat #92

THE DESERT TOWER THAT MAKES ITS OWN SHADE

In Abu Dhabi's scorching heat, engineers designed a building that **creates shade on demand** using a high-tech, nature-inspired skin.

The **Al Bahar Towers** are wrapped in a **responsive facade** made up of **1,036 umbrella-like panels** that open and close throughout the day—just like desert flowers responding to sunlight. Controlled by a central computer, the panels **reduce solar gain by 50%**, drastically cutting the need for air conditioning.

The mashrabiya-style design is rooted in **traditional Islamic architecture**, but enhanced with **modern materials and motors**, blending heritage with innovation to create **smart shading that moves with the sun**.

It's not just climate control—it's **adaptive architecture**, engineered to bloom in the desert.

Mind-Blowing Engineering Feat #93

THE ROAD THAT HEALS ITS OWN CRACKS

In the Netherlands, engineers have developed a revolutionary kind of asphalt that can **repair itself**—extending the life of roads without digging them up.

This **self-healing asphalt** is mixed with **tiny steel fibers** that allow the road to be **inductively heated** using special mobile machines. When the fibers are warmed, the surrounding bitumen softens and **fills in microcracks**, sealing the surface and **preventing potholes** before they form.

Not only does this dramatically reduce maintenance costs and traffic disruptions, it also cuts down on raw materials and carbon emissions over the road's lifetime.

It's not just a surface—it's **smart pavement**, engineered to fix itself while you drive.

Mind-Blowing Engineering Feat #94

THE SKYSCRAPER WITH A BUILT-IN FOREST

In Milan, Italy, engineers didn't just build tall—they built green. The **Bosco Verticale**, or "Vertical Forest," is a pair of residential towers that house **over 900 trees and 20,000 plants** across their facades.

But this isn't just landscaping—it's structural. Engineers had to account for **the weight, water needs, and wind resistance** of a literal forest growing on balconies. Specialized irrigation systems, reinforced concrete slabs, and wind-load testing made it possible to grow a woodland **hundreds of feet above ground**.

The vegetation absorbs CO_2, produces oxygen, dampens noise, and reduces heat—turning the towers into **living ecosystems** that change with the seasons.

It's not just vertical living—it's **nature re-engineered into the skyline**.

Mind-Blowing Engineering Feat #95

THE TUNNEL THAT COOLS A WHOLE CITY

Beneath the streets of Paris lies a hidden engineering marvel: a vast network of **underground water-cooled tunnels** that **air-condition hundreds of buildings**—all without traditional AC.

This system, known as **Climespace**, draws cold water from the **Seine River**, chills it further at central plants, then **pumps it through 50 miles of insulated pipes** to cool museums, offices, hospitals, and even the Louvre.

The setup reduces energy consumption by up to **50% compared to standard air conditioning**, slashes carbon emissions, and frees up rooftop space across the city.

It's not just infrastructure—it's **invisible climate control**, engineered below ground to cool above.

Mind-Blowing Engineering Feat #96

THE SKYSCRAPER THAT CATCHES THE WIND

In Shanghai, engineers built the **Shanghai Tower** not just to rise above the clouds—but to **harness them**.

Standing at **2,073 feet**, this twisting megastructure features **270 wind turbines** embedded near its rooftop, designed to generate enough electricity to **power the building's exterior lights**. Its spiral shape isn't just for aesthetics—it **reduces wind loads by 24%**, making the building more stable and requiring **less structural material**.

Combined with **rainwater harvesting, geothermal heating, and double-skin insulation**, the tower uses **less than half the energy** of a traditional skyscraper its size.

It's not just a marvel of height—it's a **wind-harvesting giant**, engineered to turn the breeze into power.

Mind-Blowing Engineering Feat #97

THE ROBOT SWARM THAT BUILDS WALLS

Engineers have developed a fleet of small construction robots that work **like termites**—collaboratively, autonomously, and without a blueprint in hand.

Inspired by nature, these robots use **simple local rules** and **environmental cues** to lay bricks, stack blocks, and even 3D print walls—**without a central controller**. If one robot breaks down, the rest adjust and continue the build, just like a biological colony would.

Developed by teams at **Harvard's Wyss Institute** and beyond, this swarm-based approach allows for **rapid, decentralized construction** in dangerous or remote environments—from disaster zones to Mars.

It's not just robotics—it's **emergent architecture**, engineered to build itself from the ground up.

Mind-Blowing Engineering Feat #98

THE BRIDGE THAT LIFTS LIKE A TRANSFORMER

In the UK town of Hull, engineers built a pedestrian bridge with a twist—it **lifts vertically and folds in half** like a colossal Transformer.

Called the **Murdoch's Connection**, this futuristic bridge doesn't swing or tilt like traditional drawbridges. Instead, it features a **hydraulic mechanism** that allows the entire center span to **rise straight up**, then **fold neatly into itself**, creating clearance for boats while occupying minimal overhead space.

The design combines **sleek aesthetics**, **dynamic motion**, and **engineering precision**, making it both a functional crossing and a kinetic sculpture.

It's not just a bridge—it's a **folding feat of motion and metal**, engineered for flow on land and water.

Mind-Blowing Engineering Feat #99

THE FARM THAT FLOATS ON THE SEA

In the port of Rotterdam, engineers launched a bold experiment in sustainable agriculture: a **floating dairy farm** that produces milk right on the water.

Built on a three-level barge, this high-tech farm houses **40 cows**, complete with **automated milking machines**, **robotic feeders**, and a **manure-to-fertilizer system**—all powered by **solar panels and rainwater collection**. The structure is designed to **rise and fall with the tides**, making it resilient to sea level changes.

It reduces transport emissions, recycles waste, and brings food production **closer to urban centers**, all while using minimal land.

It's not just a farm—it's a **buoyant barnyard**, engineered for the climate-challenged cities of tomorrow.

Mind-Blowing Engineering Feat #100

THE WALL THAT TRACKS SPACE JUNK

High above Earth, thousands of pieces of space debris zip around at **17,000 mph**, threatening satellites, astronauts, and entire missions. To protect against disaster, engineers built a **radar wall that sees the unseen.**

Located in **New Mexico**, the **Space Fence** is a next-generation radar system that uses **S-band technology** to track objects as small as a **screw** in low Earth orbit. Operating 24/7, it scans the sky with **incredible precision**, mapping the movement of **hundreds of thousands of debris fragments** in real time.

This information is critical for collision avoidance, satellite launches, and the safety of the International Space Station.

It's not just defense—it's a **digital shield in the sky**, engineered to watch over our place in the cosmos.

CONCLUSION

Congratulations! You've just explored *100 Mind-Blowing Engineering Feats* and taken a journey through some of the most astonishing, creative, and boundary-pushing projects ever built. From floating farms to spinning skyscrapers, underwater labs to roads that charge your car, this collection proves that engineering is far more than equations and blueprints—it's a celebration of imagination made real.

But here's the thing about engineering—it never stops evolving. For every story you've read, there are thousands more unfolding right now, each one pushing the limits of what we can build, solve, and dream up next. Maybe this book sparked a deeper appreciation for the world around you, or maybe it inspired a whole new curiosity about how things work—and how they *could* work.

Because here's the truth: engineering isn't just about structures—it's about solving problems, challenging limits, and turning ideas into reality. And the most mind-blowing part? We're just getting started.

So as you close this book, don't think of it as

the end. Think of it as the foundation—one that might just inspire the next great feat. Because the world's most incredible innovations often begin with a simple question: "What if?"

Until next time, stay curious, keep imagining, and never stop building what's possible..

ACKNOWLEDGEMENTS

Creating *100 Mind-Blowing Engineering Feats* has been a journey of curiosity, creativity, and a whole lot of "Wait, they built *what*?" While my name may be on the cover, this book wouldn't exist without the incredible minds and moments that inspired every page.

First, a huge thank you to the engineers, inventors, builders, and dreamers—past and present—who dared to defy gravity, bend materials, and challenge what's possible. Your boldness, brilliance, and sheer determination are the beating heart of this book. Every feat in these pages stands on your shoulders.

To my friends and family—thank you for putting up with my endless nerding out over tunnels, towers, and turbines. Your support (and occasional raised eyebrows) made the process that much more fun.

To the readers—thank you. Whether you picked up this book out of fascination, inspiration, or pure wonder, I hope it sparked something in you. Maybe even your own idea that the world hasn't seen yet.

And finally, to the world of engineering itself—thank you for showing us that limits are just starting points. For every wall, there's a way through it. For every challenge, a wild, brilliant, mind-blowing solution waiting to be built.

Here's to the feats behind us, the ideas ahead of us, and the remarkable things we've yet to imagine.

ABOUT THE AUTHOR

Felix Grayson is a storyteller at heart, driven by an insatiable curiosity for the strange, surprising, and downright mind-blowing achievements of human ingenuity. With a passion for uncovering the most unbelievable tales of innovation, Felix has crafted *100 Mind-Blowing Engineering Feats* to entertain, inspire, and spark wonder in readers of all ages.

When he's not marveling at futuristic architecture or diving deep into stories of wild inventions, Felix enjoys exploring science museums, devouring biographies of great inventors, and pondering life's most fascinating questions over a strong cup of coffee. A firm believer in the power of creativity and the magic of a well-told story, Felix invites you to take this journey through engineering's most incredible feats—proving that the world we build can be just as surprising as the world we live in.